Nathalie Boucher
François Bellemare

Photosystem-based biosensors for water biomonitoring

Nathalie Boucher
François Bellemare

Photosystem-based biosensors for water biomonitoring

Volume I

ScienciaScripts

Imprint

Any brand names and product names mentioned in this book are subject to trademark, brand or patent protection and are trademarks or registered trademarks of their respective holders. The use of brand names, product names, common names, trade names, product descriptions etc. even without a particular marking in this work is in no way to be construed to mean that such names may be regarded as unrestricted in respect of trademark and brand protection legislation and could thus be used by anyone.

Cover image: www.ingimage.com

This book is a translation from the original published under ISBN 978-620-2-27073-1.

Publisher:
Sciencia Scripts
is a trademark of
Dodo Books Indian Ocean Ltd. and OmniScriptum S.R.L publishing group

120 High Road, East Finchley, London, N2 9ED, United Kingdom
Str. Armeneasca 28/1, office 1, Chisinau MD-2012, Republic of Moldova, Europe
Printed at: see last page
ISBN: 978-620-5-76304-9

Table of contents

Foreword

Significant progress in analysis and treatment has been made since the 1960s, when the first observations of the effects of chemicals on aquatic life were documented: early mortalities, cancers, malformations, species extinctions, transgenic effects, etc. Similar effects have been observed in humans, including workers and the surrounding populations of major users in all commercial and industrial sectors. Similar effects have been observed in humans, including workers and the surrounding populations of large users in all commercial and industrial sectors. Why is this? Simply because chemicals of all kinds including drugs and their derivatives, hormones, pesticides applied to animals, plants and humans are increasingly added to the usual contaminants in water. These contaminants are transported across the planet via clouds, rivers, lakes, seas and oceans and are found in the food we eat. The principle of dilution that used to prevail has reached its paroxysm, water does not regenerate itself! We must take care of it to ensure its quality and its safe use. We have reached a turning point and the use of water quality control tools now exceeds the criteria provided by physical-chemical tests. Sporadic testing for targeted chemical contaminants is no longer sufficient. A test is needed that allows for a more general or comprehensive measure of water quality deterioration.

Regular monitoring of water quality not only provides characterization of its evolution, but also allows for real-time warning in case of a specific incident, alerting the operators of water treatment plants. The precursors who addressed this issue of water quality in the 1970s developed criteria based on toxicity tests using living organisms. Social and legal pressures have led to the transposition of toxicity tests on large organisms to cells and even single-cell organisms. Thus, water tests now use daphnia, algae, pimphala, fluorescent bacteria, and some large organisms such as small trout and other small fish species, etc. These tests have been called microbial tests. These tests have been called microbiotests (Persoone et al., 2000) or micro-toxicity tests (Blaise, 2000). These tests are effective, each one being complementary and, to get a global view of the water quality, several toxicity micro-tests must be carried out simultaneously. However, none of these micro-tests use the activity of photosynthetic systems, even though photosynthesis is a key mechanism essential to terrestrial and aquatic plant life and thus the basis of the food chain.

The technology presented in this paper is a biosensor responding to the presence of chemical contaminants, based on the evaluation of the activity of photosynthetic systems. The purpose of this paper is to democratize the use of the biosensor as a tool for monitoring water quality of any kind, the technology being based on well-documented scientific principles that have allowed the development of a simple protocol. The characteristics of this biosensor, previously known in scientific and commercial publications as LuminoTox, are described. Protocols and applications in aqueous media and sediments are presented in Volume II.

The matrix monitored with this biosensor is water. Why does water become a problem in the ecosystem in which life evolves? What is the importance of having a constant quantity and quality of water for the well-being of living organisms? Could biomonitoring be a way to ensure this? These topics are discussed in the first sections of the paper. The scientific principles behind this biosensor will follow later. The description of photosynthetic membranes, the functioning of electron transport and the mode of energy dissipation through fluorescence that reflects the integrity of the thylakoids will be presented in detail.

We hope that the reading of this document will bring a clearer vision of the importance of water biomonitoring and will incite a new look on the scope of water toxicity monitoring. Our objective is to propose an approach for the fabrication and use of a biosensor as a watchful eye monitoring water quality, hence the figure on the cover page.

 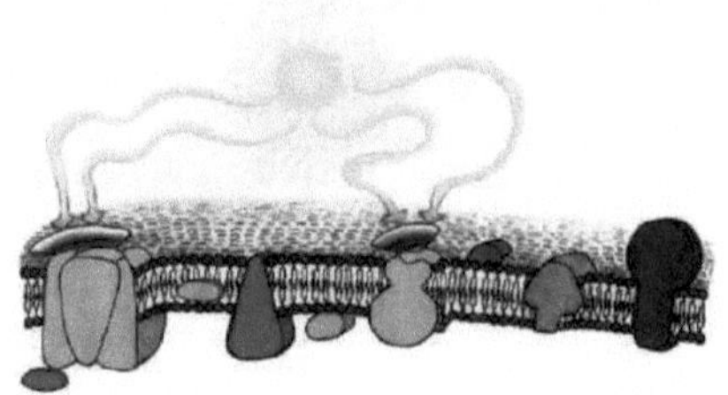

1 Water

Water is an essential substance for life. Moreover, due to its chemical characteristics, water is considered one of the best universal solvents, as well as being a mild reagent, structuring agent and carrier. This is why it is also universally used. Water quickly picks up everything in its path: any molecule, pollutant or product used ends up in water: these substances are solubilized, dispersed, aggregated, transported etc. Some of these substances are insidious because, even in small quantities, they are considered harmful to health and ecosystems. Moreover, they accumulate continuously because the treatments are not able to remove them completely. The deterioration of water quality comes from different sources, both natural and human, as a result of sports, urban, industrial and agricultural activities (Figure 1).

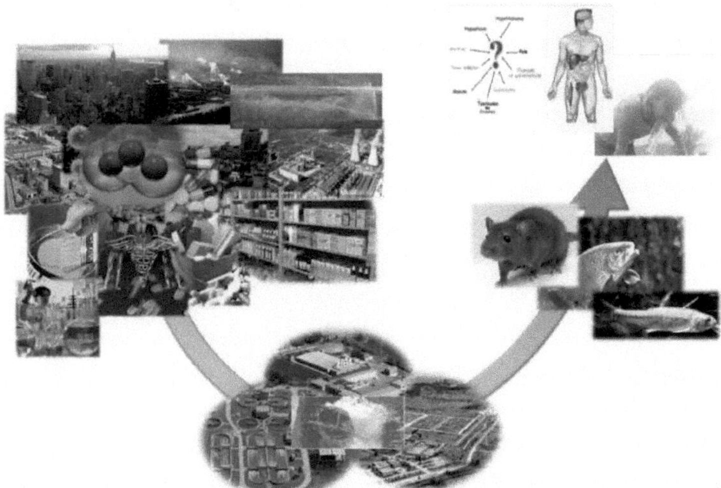

Figure 1: Water contamination (Courtesy of Dr. Robert Hausler, ETS, Montreal, Canada). The chemical contaminant is represented by the three red balls. The origin of the contaminants is both domestic, agricultural and industrial. The effects are equally felt on terrestrial and aquatic fauna and flora. In humans, several different organs can be affected by pollutants.

In order to ensure high quality water, regular physicochemical tests such as pH, color, odor, suspended solids, turbidity, dissolved oxygen are simple tests that should be performed on any water sample. In addition to these basic physico-chemical characteristics, the presence of microorganisms and chemical contaminants in quality and quantity also defines water quality. Some of these analyses are however more laborious and must be performed in specialized laboratories. Afterwards, appropriate treatments are applied to the water in order to remove all contaminants, whether bacterial, viral or chemical. These treatments range from the simplest, such as decantation or filtration, to the most complex, such as chemical transformation (oxidation) or agglutination (coagulant). There are costs associated with these treatments because the amount of water to be treated is increasing and, not to be forgotten, the water is constantly reused. Thus, contaminants that are not removed or transformed by the treatments, initially present in trace amounts, accumulate in quantity and quality. In addition, there is diversification of these substances when they react with each other, which is known as by-products. Chemical by-products can also be generated by the products and radiation used in the treatment processes. Given this complexity, a more comprehensive approach that includes a cumulative effects toxicity test, performed before and after treatment, has become essential. Indeed, toxicity measurement takes into account the presence of chemical contaminants acting alone or in groups, producing an additive or even synergistic response (the effect of two or more substances on a living organism is greater than the sum of the individual effects of these substances, one of which may act as an adjuvant).

1.1 Water: physico-chemical properties

Why is water both so important and so potentially problematic for environmental and human health? To explain the impressive qualities of water, let's briefly review its properties.

1.1.1 Water: its dipolar nature

Figure 2 shows the water molecule emphasizing its dipolar nature. The water molecule forms an angle of 104.45° at the oxygen atom between the two hydrogen atoms. Since oxygen has a stronger electronegativity than hydrogen, the side of the water molecule where the oxygen atom is located is negatively charged (partial charge), thus in opposition to the section occupied by the hydrogen atoms. A molecule with such a difference in charge is called dipolar. The water molecule has a dipole moment of 1.83D (Debye).

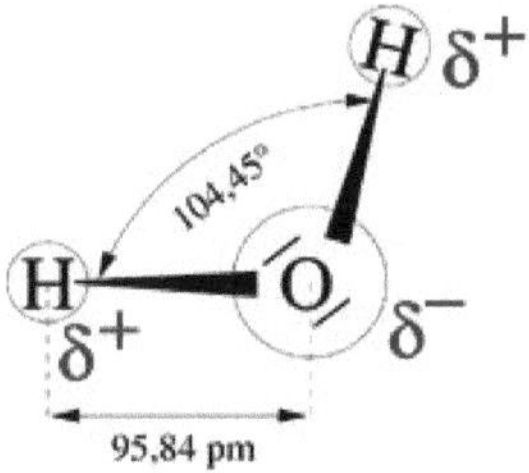

Figure 2   Typical representation of the water molecule.

(https://fr.wikipedia.org/wiki/Mol % C3%A9cule_d%27eau) (Wikipedia Commons)

This orientation of positive charges towards the hydrogen section and negative charges towards oxygen causes water molecules to behave like small magnets that attract each other, with the positive side of one attracting the negative side of another. This dipolar property allows water to appear in a variety of organized structures in the liquid and solid state.

1.1.2 Water : a network of hydrogen bridges

Bridges or hydrogen bonds are intermolecular bonds whose constituent atoms respect an orientation so that in a network of H-bridges, a molecular architecture is optimized. The simplest example is that of water molecules whose ordered structure and physical-chemical properties depend on these bridges. These bridges are woven between a hydrogen constituting a water molecule and the oxygen constituting the neighboring water molecule and so on in order to link all the water molecules together and thus form a network (Figure 3). It is an intermolecular force of attraction between the bound hydrogen (electropositive) and the bound oxygen (electronegative). The strength of this bond is equivalent to 10% of that of the covalent bond.

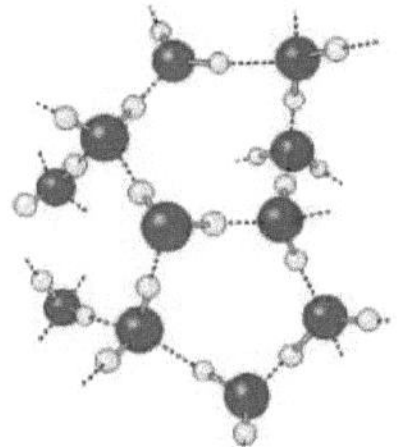

Figure 3   Representation of the hydrogen bond.

(https://fr.wikipedia.org/wiki/Liaison_hydrog%C3%A8ne) (Wikipedia Commons)

The importance of hydrogen bridges is first apparent in the equilibrium linking the three gaseous, liquid and solid phases of water (Figure 4).

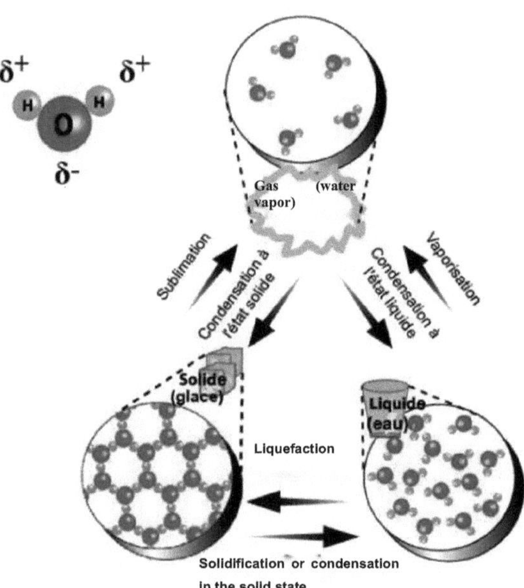

Figure 4: The three phases of water: gaseous (single molecules), liquid (molecules close together) and solid (crystal lattice). From http://ressources.fondation-

Because of its dipolar properties and hydrogen bridges, liquid water has properties that distinguish it from molecules with similar structure. Thus the density of liquid water is higher than the density of ice (the state where water is crystallized). It reaches its highest density at the temperature of 4°C, the temperature found at the bottom of a frozen pond or lake. It is at this temperature that oxygen has a greater solubility in water. Water also has a much higher boiling point than other similar molecules because more energy is needed to break the hydrogen bridges and pull the molecules apart to isolate them in the gaseous state.

1.1.3 Water: an acid and a base

Water reacts with itself so that it forms hydronium $H3O^+$ and hydroxyl OH^- ions. This is the dissociation equilibrium reaction where hydronium $H3O^+$ and hydroxyl OH^- ions are formed:

$$2H_2O \longleftrightarrow H_3O^+ + OH^-$$

The product of the concentrations of these ions, or "dissociation product", is constant. At 25°C, it is :

$$[H_3O^+]\cdot[OH^-] = 10^{-14}$$

(no unit because it is an equilibrium constant). The concentration (in mole per liter) is, by convention, symbolized by the square brackets.

Hydronium and hydroxyl ions are very reactive, they can attack various materials, transform them and even dissolve them. The acidity, the concentration of hydronium ion, is defined by a mathematical representation, the *pH*:

$pH = -\log_{10}[H_3 O^+]$. As in the case of water, $H_3 O^+$ and OH^- are intimately related, we define $pOH = -\log_{10}[OH^-]$ and because of the equilibrium, $pH + pOH = 14$

At 25°C, the *pH* of pure water is defined as being 7; it is said to be neutral. The addition of certain so-called "acidic" products will shift the dissociation equilibrium of the water and lower the *pH* (because the product of the two concentrations must remain $[H_3O^+]-[OH^-] = 10^{-14}$. If there is an increase in the amount of hydronium ions, they will react with the hydroxyl ions and the medium becomes acidic. Vice versa, the addition of certain "basic" products will induce the reaction in the opposite direction, i.e. promote the presence of hydroxyl ions, decrease the concentration of hydronium ions and thus increase the *pH*. We note that water in its dissociation reaction can capture a proton or release one, it is an ampholyte, ie can act as an acid or a base.

1.1.4 Water: a solvent

Because of its polarity, water is an excellent solvent. When a salt or a polar compound comes in contact with water, it interacts with the water molecules. The relatively small size of water molecules means that many of them surround the solute molecule. The negative dipoles of water attract the positively charged regions of the solute and vice versa. In general, ionic and polar substances such as acids, alcohols, sugars and salts dissolve easily in water, and non-polar substances such as oils and fats dissolve with difficulty. However, these non-polar substances aggregate and form a separate phase in the aqueous medium, thus becoming a non-polar portion suspended in water. The water molecule also hydrates inorganic ionic substances (metallic, non-metallic and radicals), but also organic ions. Water thus forces the ionic molecules to structure themselves, for example in micelles for the largest ones, to remain in suspension. By this bipolar property, water can be loaded with ionic substances, polar and non-polar substances, which, in turn, will organize themselves into hydrated structures, dispersed or suspended in water. These colloids adsorb any organic or non-organic chemical substance, whether ionic, polar or non-polar, simple or complex.

1.1.5 Water: surface tension

The dipolar attraction between the water molecules gives it a high cohesion and therefore a high surface tension. This can be seen when small amounts of water are placed on a non-soluble surface and the water molecules remain together in the form of droplets. This property, which is manifested by capillarity, is useful in the vertical transport of water in plants.

1.1.6 Water : carrier of life

As long as there has been no water there has been no life, which is why water is both so important and, indeed, so potentially problematic for human and ecosystem health. These properties of water are exploited to the maximum in the living and even in its emergence. It is the essential molecule. Spring water is filled with living organisms and the ecosystem created is fragile. Aquatic organisms feed on what they find in the water. Their survival and that of their predators in the food chain is dependent on the quality of the water. If molecules harmful to their metabolism are present, the organisms die or migrate to a more hospitable site. This is the case for higher species that drink and feed on organisms that have lived in more or less favourable or even polluted environments. Thus, contaminants are transmitted through the food chain (Figure 1).

1.1.7 Water: recurrent contamination

The presence of contaminants in water is evaluated by microbiological and physico-chemical unit analyses and their quantification in water must be below a standardized quantity as being the concentration not to be exceeded (acute or chronic toxicity index). For chemical contaminants, analyses are time consuming and expensive, requiring advanced technologies and performed by specialized laboratories. This results in a lack of interest in performing these analyses and consequently, a lack in monitoring water quality. Given the multitude of chemicals that may be present in concentrations exceeding the standards, only a few typical molecules are actually analyzed. The list of contaminants is regularly revised, but in the meantime, new problematic molecules (e.g., nanomaterials, endocrine disruptors, etc.) arrive rapidly and are ignored by regulatory agencies. In addition, potential synergistic effects are excluded from the standards. For example, if lead is suspected in water, only lead will be chemically measured, but if it is accompanied by cadmium or chromium, its toxicity to living organisms is increased, and thus the toxicity level of the mixture for humans. This toxicity will not be assessed chemically, which is inappropriate for water quality monitoring.

It is because of the effects of chemical contaminants on living organisms that biologists established in the 1960s and 1970s a new branch of biology, **ecotoxicology**. This science can be defined by the whole of any substance or preparation which presents or can present immediate or deferred risks for one or more components of the environment. It is also the combination of two sciences, ecology and toxicology. Until now, studies in environmental toxicology were mainly concerned with the harmful effects of toxic substances on humans. Ecotoxicology thus takes into account the effects of chemicals in the context of ecology, i.e. the adaptive capacity of the organisms present.

2. Water biomonitoring

The deployment of the science called ecotoxicology, in terms of its growing complexity, has led to the interest of scientists of all backgrounds. Applied in the water sciences, tools for measuring ecotoxicological response have developed and have become more meaningful than measuring the presence of one or more contaminants in water. It is a look at the impacts of water quality on living organisms (Figure 5). Ecotoxicological measurements thus allow for the monitoring of water by measuring the improvement or degradation of its quality. Biomonitoring therefore depends on the processes put in place to verify and follow the evolution of the measured parameters.

Until the early 1990s, water quality standards were based on physicochemical and microbiological parameters. However, this approach proved to be limited because water quality also depends on the interaction of all these elements together. Since then, the concept of biomonitoring has been developed. When a stream, effluent or any other water sample is biomonitored, this implies that the impact of chemical contaminants is integrated into the conventional physico-chemical monitoring parameters. This is done by measuring a living organism that is well adapted to the water and that responds appropriately (favorably or unfavorably) to the environment or water sample. Biomonitoring can then be seen as a lookout (Figure 5) for unwanted substances. Expressing this concept to water contamination, the eye that analyzes the water becomes a spotter (an operator), the chemical contamination represents the undesirable unknowns, and the eye through which the water is monitored is the test that indicates the presence of these unknowns. If chemical contamination increases, water quality decreases, and the impact on the health of aquatic organisms, and consequently on humans, increases. Biomonitoring therefore means the evaluation of the impact (toxic effect) of chemical contamination on a selected living organism, under well-defined conditions (Keddy et al. 1994).

Microbiotests or micro-toxicity tests, which have been developed since the 1975's, allowed the first biomonitoring analyses of water. The pioneers in this field were able to convince the competent authorities of the accuracy of the results and their more global vision than conventional physico-chemical tests including unit measurements. Toxicity micro-tests use different physiological parameters to measure the toxicity of water. In addition, they are simple to use and inexpensive, so they can be used on a regular basis as a water quality monitoring tool in the same way as standard physicochemical tests (Persoone et al., 2000; Blaise et al., 2005).

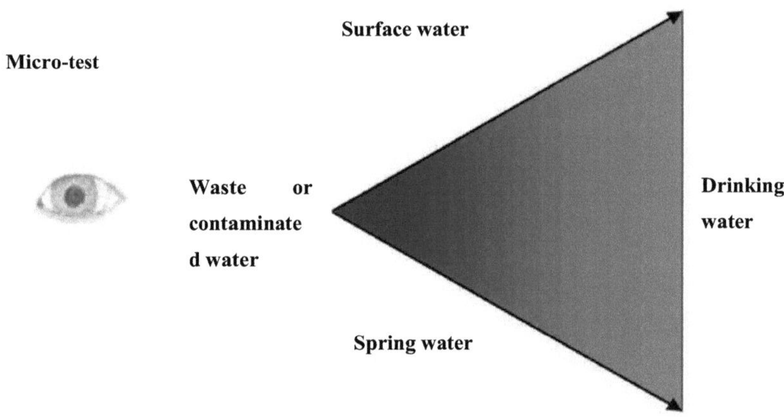

Figure 5: Water quality lookout, the toxicity micro-test or water quality eye.

2.1 Ecotoxicological response assessment: bioassays, microbiotests and toxicity microtests.

The application protocols for ecotoxicological tests are based on conventional toxicity tests. Because of the potential complexity of the "water" matrix, a multitude of chemical substances are taken into account by these tests. They therefore provide a more global and representative response to the environment than an analysis of a single substance as performed by toxicological studies.

As mentioned earlier, an ecotoxicological test is an experimental test that determines the effect of one or more substances on a selected group of organisms under well-defined conditions (Keddy et al, 1994). These tests use different means to measure the toxicity of a product. The most commonly used means is the measurement of mortality or reproduction, but there is a growing interest in using more sensitive endpoints. Biochemical, physiological, reproductive and behavioral effects can also be indicators of the presence of toxic molecules. Most ecotoxicity tests provide an estimate of the dose that affects 50% of the population. This may be, for example, the average lethal concentration that kills 50% of the population. It is also possible to estimate the maximum concentration that causes no effect. Many terms related to toxicity testing have been defined. The most common are the LD50, which is the dose of toxic pollutant that produces a 50% lethality compared to controls. If tests are performed on "determinants" other than mortality, for example on weight or energy reserves, then an EC50 (effective concentration) is defined. The EC50 is the concentration of pollutants affecting 50% of the population or of the parameter measured. In addition, depending on the experimental design and the concentrations of toxic pollutant chosen, other values can be used, such as the ECx (with x=20% for example, representing the concentration having a 20% effect on the selected activity of the population). It is from these parameters defining ecotoxicology that microbiotests, toxicity microtests, biosensors and even biosensors have been developed.

These ecotoxicity tests applied for water quality monitoring, or toxicity micro-tests, can be distinguished into two types:

Acute toxicity tests: they are carried out over a very short period of time (compared to the generation time of the organism). Their advantages are their speed and low cost. These tests generally involve high concentrations of the pollutant; as a result, the long-term effects of low concentrations are not highlighted. One example is the

24-hour daphnia immobilization test. (For an exhaustive review of toxicity microtests, see Persoone et al., 2000 and Blaise & Férard, 2005).

Chronic toxicity tests: they are carried out over a relatively long period of time compared to the generation time of the organism. These are, for example, tests on reproduction. They are longer and more costly than acute tests, but they allow the long-term effects of a pollutant to be highlighted[1] .

---

1 *Pollutant*

Definition: Physical, chemical or biological agent present in an ecosystem, which causes by its concentration, disturbances prejudicial to its good balance and reduces the possibilities of its use.

3 Biosensor based on photosynthetic systems as a toxicity microtest

The use of living organisms in ecotoxicity tests makes the test procedure long, laborious and subject to more or less objective criticisms, which sow doubts in the observer's mind (Blaise, 2000; Cairns, 2005). It is necessary to ensure the adequate maintenance of living organisms in order to maintain their characteristics in order to have a reliable and reproducible test. The test protocol must be carried out under well-defined and reproducible conditions, by well-trained and dedicated personnel. To overcome these difficulties, simple, inexpensive methods capable of rapidly detecting the toxic effects of an industrial or agricultural effluent or simply of a chemical substance have been developed (Isomaa et al., 1995). An assay that measures enzymatic activity rather than cell growth thus allows the use of metabolic constituents, or even organelles, and thus simplifies manipulations that respond in real time to several types of molecules.

2.2 Biosensor: presentation of biological material made from photosynthetic systems

By convention, a biosensor is defined as a device consisting of three parts:

- **Biological material** (enzymes, antibodies, tissue, receptors, organelles, nucleic acids etc.) that responds to one or more specific molecules;

- **The transformation of the biological signal of the presence of the target molecule(s) or pollutant into a signal detectable by a device;**

- **The transducer or reader sensitive to the biological signal** (optical, amperometric, potentiometric, thermal etc.).

The biosensor explained in this paper uses chlorophyll photosystems of plants, algae and cyanobacteria as biological material. Its advantages are the availability of the material and the fundamental evolutionary knowledge of its physiology whether in the structure or the functioning of chloroplasts, thylakoids or photosystems and the possibility to obtain genetically modified or even synthetic structures. For all these reasons it is an ideal material for the realization of a biosensor or micro-toxicity test. Indeed, many extrinsic molecules of all chemical and biochemical families subtly disrupt its functioning and these effects are reflected in the measurement of the activity of photosystems.

3.1.1 Structure of thylakoids and photosystems of plants, algae and cyanobacteria.

In higher plants and some algae, photosynthesis takes place in specialized organelles, the chloroplasts. The chloroplast is a quasi-cell with its genetic code and mitochondria inside the plant cell. The photosynthetic reaction takes place in specialized structures of the chloroplast, a membrane system called thylakoids. They are located in the stroma, the aqueous compartment of the chloroplast.

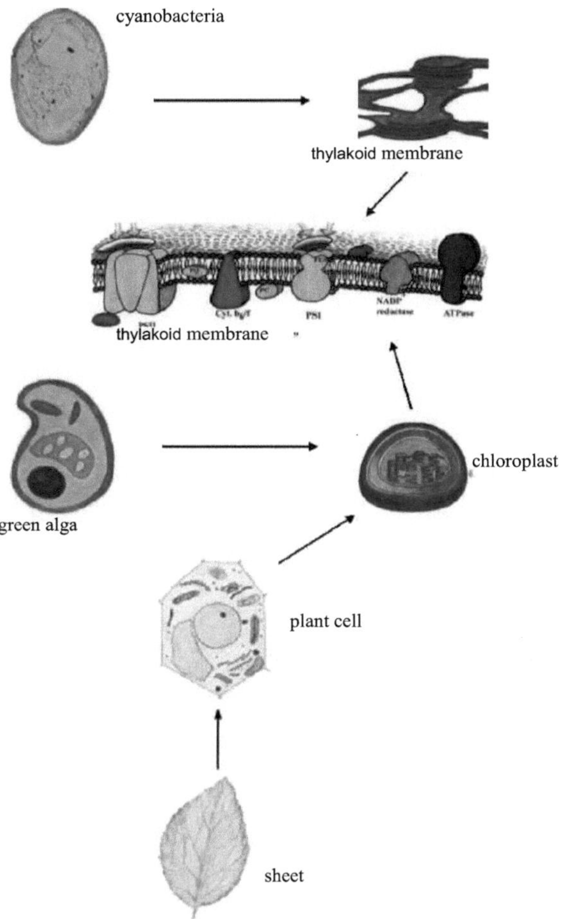

Figure 6: Conservation of the thylakoid membrane structure of the cyanobacterium at the upper plant (Illustrated montage by Nathalie and Isabelle Boucher and Louis Bellemare).

Figure 6 illustrates this evolutionary nature of thylakoids, loose membranes loosely stacked in cyanobacteria then increasingly structured in chloroplasts of algae and plant cells.

Some species of algae and cyanobacteria, do not necessarily contain chloroplasts, however they do have thylakoids, structures similar to those of higher plants (Neilson et al., 2010). It is increasingly recognized that chloroplasts are likely derived from a common ancestor with a membrane arrangement of thylakoids, which is found in cyanobacteria.

In higher plants and algae, thylakoids form a network consisting of folded membranes and form semi-cylindrical structures, 5 to 20 tiers or discs named grana Figure 7A). They are connected to each other by single membranes, the lamellae (Figure 7A, Pfeil et al. 2014). The outer surface of the thylakoids is immersed in the stroma while the inner side surrounds an aqueous compartment, the lumen. The lumen constitutes the space between the thylakoids, a space that is continuous between the folded regions and the lamellae (Foyer 1984). In cyanobacteria, the presence of a huge light collector, the phycobilisomes (Figure 11), prevents the layered arrangement of thylakoid membranes. In this case, these stacks are more modest, in groups of 2 or 3. Note that these foldings are found in groups of 2 or 3 in diatoms and brown algae, which do not have phycobilisomes, suggesting an evolutionary sequence in the structure of thylakoids.

Thylakoid membranes contain the photosynthetic apparatus of the chloroplast and are the organelle responsible for the functioning of chlorophyll-based photosystems. They consist of important protein complexes characterized by the antenna complexes of photosystems I and II (PS I and PS II), the reaction centers of PSI and PSII, the cytochrome b complex$_6$ /f, and the ATP synthetase complex (Figures 7B and 8, Murphy, 1986). They are distributed in the membranes as suggested by the illustration in Figure 7 B and detailed in Figure 12.

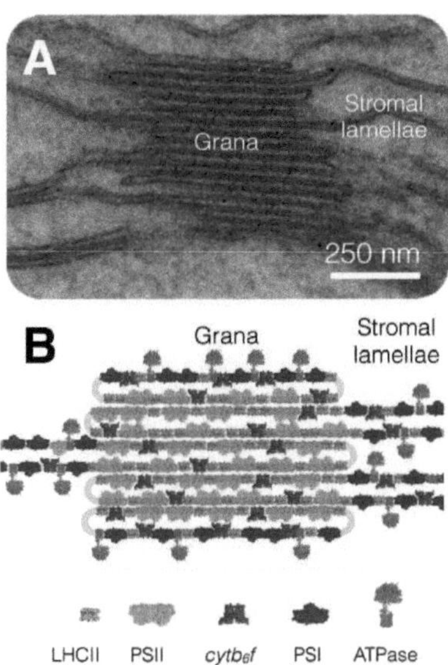

Figure 7 A: Grana structure in chloroplast thylakoid membranes. The stacked and lamellar membrane structures are shown and. **B: Distribution of thylakoid enzyme complexes (as indicated by the symbols representing them).**

(Courtesy of Portland Press as published in: Johnson, M.P., Photosynthesis, Essays in Biochemistry 201;6 60 255-273.

The particular organization of thylakoid membranes favors the optimal activity of both photosystems I and II, lipoprotein complexes in which photochemical reactions take place. However, for the measurement of the fluorescence related to the photochemical reactions of the photosystems, the main actor is PSII, detailed in the next section. Also, as it is the antenna complexes that capture the photons and direct them to the reaction centers of the photosystems, we describe below some of these antenna complexes as they are an integral part of the suggested biomaterials for the biosensor: higher plants, algae and cyanobacteria.

Preparations of thylakoid membranes contain the entire photosynthetic system.

By conventional gentle methods, which range from the splitting of plant cells and chloroplasts, it is possible to extract these membranes and make vesicles from them which retain the physiological properties of the

photosystems. Thus, in addition to living cells, algae and cyanobacteria, the
prepared membranes of thylakoids constitute excellent photosynthetic
materials with complementary properties. We have not yet exploited the full potential
of the biomaterial that are photosynthetic systems for
example measurements via the cytochrome $b6/f$
complex or ATPase.

3.1.2 The photosystem II (PS II)

The PSII, inserted in the stacked portion of the granum, (Figure 7) plays an essential
role in electron transport. Indeed, it is responsible for the photolysis of water i.e. the
separation of charges into its constitutional elements. Oxygen is released into the
atmosphere while hydrogen is used as a reductant of CO_2 to produce carbohydrates
(Barber, 1993). In addition, PSII functions to store energy by separating and
stabilizing the positive, $H+$ protons, and negative charges on either side of the
thylakoid membrane. Because of its functions, it is possible to determine the
photosynthetic activity of PSII by measuring oxygen release, energy storage or
fluorescence emission.

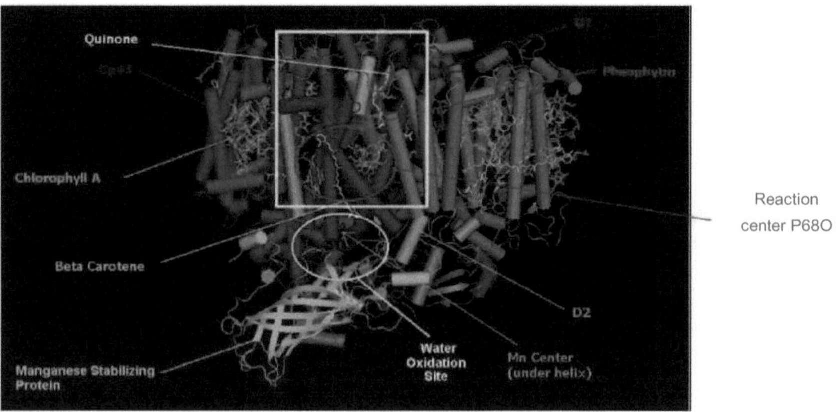

Figure 8:   Schematic and structure of the PSII of a plant cell. The quinone, as shown in this figure refers
to Q_A in the text. As for Q_B , it is the binding site on photosystem II of a quinone from the quinone pool. The

rectangle locates the intermediates involved in electron transfer (detailed in Table 1). This figure by Neveu, Curtis (C31 004) is from:https://upload.wikimedia.org/wikipedia/commons/thumb/a/a1/Photosystem-II_2AXT.PNG/220px-Photosystem-II_2AXT.PNG The P680 reaction center.

PSII as illustrated in the dynamic diagram in Figure 9, consists of a heterodimer formed by 2 transmembrane proteins D1 and D2, of 32 and 34 kDa respectively, which stabilize the pigments and electron transporters, quinones of the reaction center (Ghanotakis et al., 1990). These two proteins alone bind all the cofactors as well as the constituents necessary for the primary light energy conversion process (Barber, 1993). These components of the reaction center, are tyrosine Z, P680, pheophytin (Ph) and plastoquinones Q_A and Q_B. All these molecules play a major role in the photochemical mechanism of charge transfer.

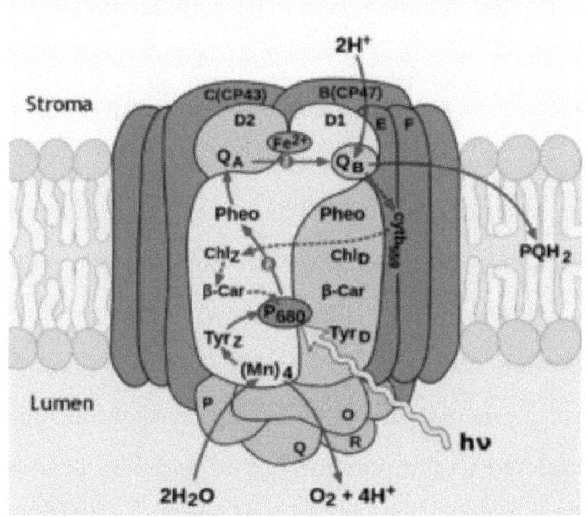

Figure 9:   Operation of the PSII reaction center. The pathway of electrons from water hydrolysis to Q_A and Q_B is represented by red arrows. This figure by Kaidor is from https://Commons.wikipedia.org/wiki/File:Photosystem_II_-_ru.svg.

The P680, is the reaction center formed by a pair of chlorophylls. It is the photochemical center which receives the energy of the photons absorbed and channeled by the antennas. The photochemical reaction is summarized in a separation of charges and their transfer to quinones. Without this charge separation, no photosynthetic activity would occur. This reaction center is designated P680 because it absorbs maximally at a wavelength of 680 nm. We also denote in its environment the presence of cytochrome b_{559}, a protein consisting of two sub-units of 10 and 4 kDa.

Extrinsic PSII polypeptides of 33, 23, and 17 kDa are localized on the inner surface of the thylakoids. These polypeptides are part of the oxygen release complex. Several inorganic or mineral ions (Mn, Cl, Ca, Fe and CO_3^-) are also involved in the oxygen release process. It is triggered when tyrosine Z is oxidized (because it has given up an electron to P_{680+}). As schematized in Table 1, the reduction of z_+ occurs 4 times in order to store the appropriate number of oxidative equivalents and split the water to produce oxygen molecules (Barber, 1993).

3.1.2.1 Antenna complexes in plants, algae and cyanobacteria

Photosynthesis is triggered when photosystems receive light at their reaction center and induce a charge separation by photochemical reaction. To regulate their light absorption, photosynthetic organisms have developed different antenna patterns consisting of photon-absorbing pigments at various wavelengths from ultraviolet to infrared. The important constituent proteins of the antennae participate in the optimized arrangement of the pigment-protein complexes. This orderly and delicate arrangement is essential for efficient transfer of photon energy from the sensors to the reaction center, of which P680 is a limiting step (see review by Croce and Amerongen, 2014). Its role is also that of an energy dissipator (Neilson et al., 2010). It is a finely orchestrated arrangement of pigments, chlorophylls, and carotenoids that channel solar energy to transmit it regulated to the reaction center and also acts as protectors against collateral oxidative reagents produced, which Havaux et al. (1999) refer to as the excessive light protection system. Chloroplast genes allow the genetic transmission of this achievement.

Eukaryotes acquired photosynthetic metabolism over 1 billion years ago. During this time the light collecting antennae have undergone structural and functional divergence. Phycobilisomes (PBS) formed the first antenna sensors in photoautotrophs and these protein complexes continue to be dominant in cyanobacteria, glaucophytes and red algae. The PBSs have been replaced by the light sensing antenna complexes (LHCs) found in the majority of eukaryotes.

Let us briefly describe some sensor antennae. The three main ones: the internal antennas found in the reaction centers of PS II and PS I, the external antennas or LHC and the phycobilisomes (PBS) absorbing pigments of cyanobacteria.

3.1.2.1.1 LHCs (light collecting complexes)

Two families of pigments are found in light-sensing complexes forming so-called antennae. These are tetrapyrroles and carotenoids that come from the biosynthesis of isoprenoids. There are cyclic tetrapyrroles, chlorophylls and those with an open chain, phycobilins. Among the chlorophylls, there is chlorophyll *a* (Chl *a*) which is the main pigment of the reaction centers (RC) and light collecting complexes (LHC) of green algae and higher plants. To broaden the range of absorbed photons, Chl *a is* assisted by other chlorophylls and carotenoids, called accessory pigments. The absorption spectra are shown in Figure 10. These pigment arrangements are the main ones found in the antenna complexes. The antenna complexes in the reaction centers, also called internal antennae, are integrated or even fused with the proteins that also carry the cofactors for charge separation, which is an integral part of the structure and function of the reaction centers. These structures are found in the internal antennae of PSII, PsbB (CP47) and PsbC (CP43) (Figures 8 and 9) and in the PSI of higher plants, chlorophycean algae and cyanobacteria.

There are also the light-harvesting complexes (LHCs). These are external antennas that collect photons, found in higher plants and chlorophyceous algae.

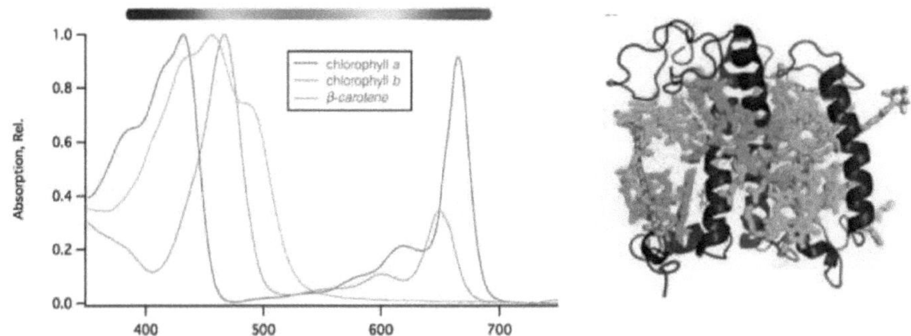

Figure 10: PS II antenna pigment complex from spinach chloroplasts (PDB 1RWT 46). Left: absorption

spectrum of this complex. Right: structural representation of this antenna complex: apoprotein (blue), Chl *a* (green) and Chl *b* (cyan). Carotenoids are presented in yellow and orange. Lipids mainly galactolipids are not represented as well as phytile chains. Croce R. and van Amerongen H. (2014). Natural strategies for photosynthetic light harvesting *Nature chemical biology*, 10, 492 -501 (with permission from Nature Publishing Group).

LHCs are composed of non-covalently bound pigments (chlorophylls and carotenoids). They are localized at the periphery of PSII and PSI and respectively named LHCII and LHCI. Plant LHCs proteins are formed by 3 trans-membrane helices that bind Chl *a*, Chl *b* (attached by their central Mg^{2+} in different ways), and some carotenoids (Figure 10).

3.1.2.1.2 PBS (phycobilisomes)

The external antennae of cyanobacteria (Figure 11) and red algae differ from those of plants and algae. Phycobilisomes (PBS) are water soluble proteins, a 5-10 Mda complex (representing 40-50% of the total proteins in the cell). These proteins lack transmembrane helices and their pigments, phycobilins, are covalently bound. The pigment composition and complex organization allow photon absorption in a broad portion of the visible spectrum. PBS organisms are among the organisms with the greatest adaptation to the quantities and qualities of incident radiation. They are megacomplexes, well settled on the PSII. PBS transfer their energy to the internal antennas of PSII and PSI.

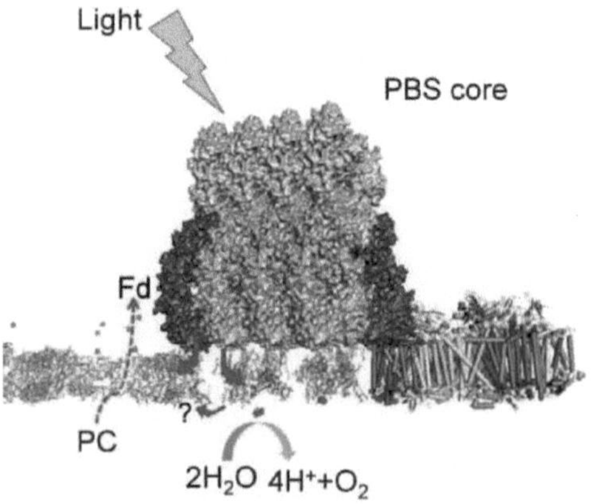

Figure 11: Modeled structure of the PBS-PSII and PBS-PSI megacomplex. PSII is completely covered

by the phicobilisome center, while PS I is indirectly covered via the ApcD moiety, an allophycocyanin.

Liu H, Zhang H, Niedzwiedzki D.M., Prado M, He G, Gross M.L. Blankenship R.E.(2013). Phycobilisomes Supply Excitations to Both Photosystems in a Megacomplex in Cyanobacteria. Science; 342: 1104-1107. With permission from AAAS.

The main phycobilins of cyanobacteria and red algae are phycocyanin, phycoerythin and allophycocyanin, whose maximum absorption wavelengths are :

- phycoerythrins (PE; Amax = 565nm)

- phycocyanins (PC; Amax = 620nm)

- allophycocyanins (APC; Amax = 650nm)

Cyanobacteria or blue-green algae preferentially absorb the red color of the visible light spectrum.

3.2 Biosensor: measurement of photosynthetic activity as an index of thylakoid integrity

The preparation of the biological material is validated by regular quality control tests by the constancy of the values obtained, under defined conditions, by the measurement of the photosynthetic activity. As photosystems are activated very quickly in the light, in the order of a millisecond, any disturbance in their conformation will be faithfully translated in the evaluation of the photosynthetic activity. It is the transformation of the biological material that becomes the physiological activity of reference of this biosensor. The integrity of the biological material is measured by the fluorescence measurement, its relation with the photosynthetic activity will be explained later. Let us first describe the light phase of photosynthesis.

3.2.1 The main steps of photosynthesis

In the general process of photosynthesis, chloroplasts use water and carbon dioxide to produce carbohydrates and oxygen in the presence of light (Figures 9 and 12). To form the carbohydrates essential to the plant, photosynthesis is distinguished into two phases, the light phase (electron transport), occurring in the thylakoid membranes, and the dark phase (Calvin cycle), occurring in the chloroplast stroma. Electron transport provides NADPH and ATP so that sugars, such as starch, can be produced by the Calvin cycle. The overall reaction of photosynthesis (reaction 1) is written as follows:

$$nH_2O + nC_2 \xrightarrow{h\nu} (CH_2O)n + nO_2 \qquad (1)$$

The membranes of the thylakoids are responsible for the light phase of photosynthesis, while the reactions of CO_2 fixation and sugar formation take place in the cytoplasm of the chloroplasts. The steps of the light phase can be summarized as follows:

- the hydrolysis of water

$$2\ H_2O \xrightarrow{h\nu} 4H^+ + 4e^- + O_2 \qquad (2)$$

and, in order to explain what happens to the electrons:

- the role of plastoquinones

$$2H_2O + 4\ \text{photons} + 2PQ + 4H^+ \longrightarrow O_2 + 4H^+ + 2\ PQH_2 \qquad (3)$$

Reactions 2 and 3, which boil down to oxygen release reactions, take place in the so-called reaction center of PS II.

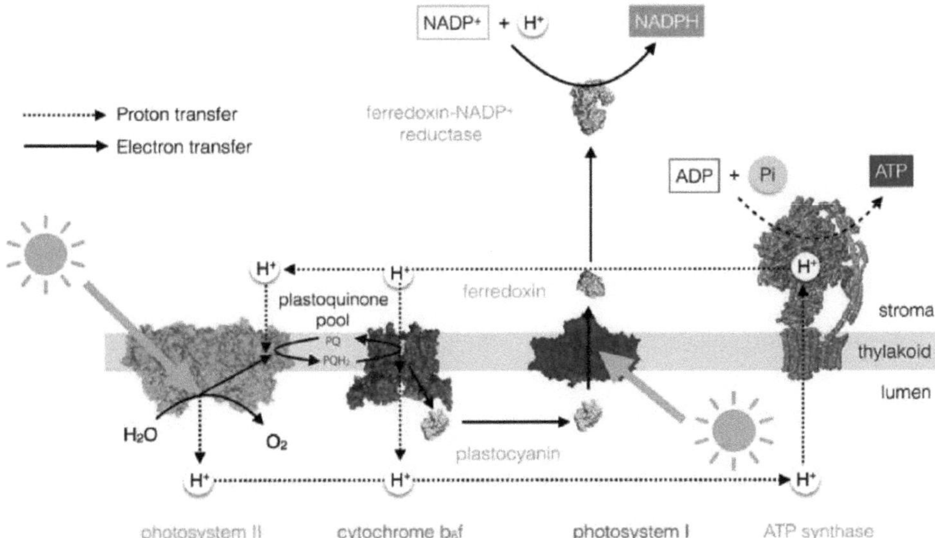

Figure 12: Light phase of photosynthesis. This is a representation of the electron and proton (H+) transfer chain in order to reduce NADP+ to NADPH and form a proton gradient on both sides of the thylakoid membrane in order to activate the ATPase.

Courtesy of Portland Press as published in: Johnson, M.P., Photosynthesis, *Essays in Biochemistry* 2016; 60 255-273.

3.2.2 Electron transport in photosystems.

Recall that the light phase is carried out by various protein complexes classified among proteins, pigments and lipids whose role is 1) to capture energy from the sun, 2) to transfer this energy to the electron transport chain and finally, 3) to hydrolyze water (Figure 12). This cascade of events is triggered by the absorption of light by the antennae of the two photosystems: PS II and PS I. In PSII, this energy is transferred to the reaction center (P680) which will then go to the excited state. To return to the ground state, an electron from P680 will quickly pass to a neighboring molecule: the pheophytin. This event is called "charge separation": the pheophytin (Ph⁻) takes the electron of P680 and will then carry a negative charge and P680+, a positive charge. The P680+, is not a stable ion, it is even a very powerful oxidant. It will take an electron from tyrosine Z (Tyr Z), which will regenerate by accepting an electron from the oxygen release complex (water oxidation cycle). The oxygen release complex will then undergo a 4-step cycle that leads to the hydrolysis of two water molecules into four electrons and 4 protons and release the oxygen (reaction 2). This reaction occurs in the reaction center via the oxygen release complex which involves manganese ions (Mn^{2+}, Mn^{3+} and Mn^{4+}) as cofactors. This O_2-producing water hydrolysis is essential, not only for the regeneration of P680 from P680+, but also for the emergence of aerobic life caused by the gradual increase of its concentration in the medium as the activity of photosynthetic organisms *proceeds*.

The sequence of reactions in Table 1 presents a model of the charge separation steps induced by photon excitation of P680. These steps result in electron transfer involving the surrounding molecules of P680: tyrozine Z, pheophytine, and quinones. These are the essential phases of the O_2, proton and electron production cycle during the light phase of photosynthesis, from the absorption of a photon to Q_B (Scheme adapted from Krause and Weis, 1991).

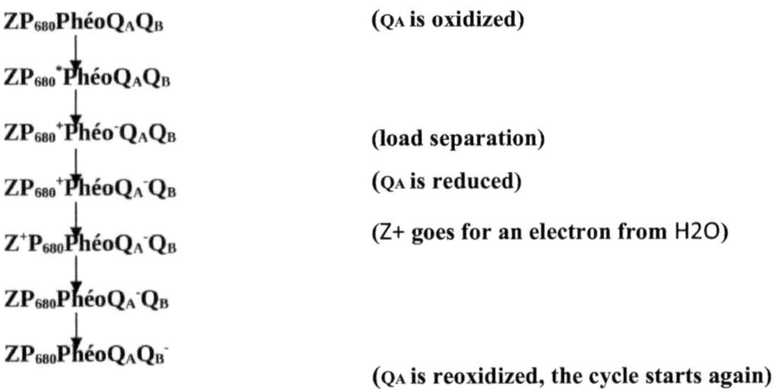

Table 1. Sequences of reactions leading to charge separation from P680.

These reactions occur in the two proteins D1 and D2 of photosystem II: Ph- transfers its excess electron to the plastoquinone, Q_A. FOR PSII to complete its functional role, the electron from Q_A is transferred to Q_B, which is also a plastoquinone. However, unlike Q_A, Q_B accepts a second electron and becomes completely reduced. For the complete reduction of PQ to form PQH_2, in addition to the electrons from Q_B, two H^+ must come from the stroma (reaction (3), as detailed in Table 1). Both plastoquinones Q_A and Q_B are retained there via their isoprene chain, localizing their polar heads near the surface. The polar head of Q_B is called upon to receive two protons to form Q_BH_2 and, as a result, the quinol ring is oriented differently so that it can present the protons to the cytochrome b complex$_6$ /f. This cytochrome acts as a catalyst in the proton transfer reaction from PQH_2 to plastocyanin (Pc) and allows their release into the grana lumen. The Fe ion^{2+} acts as a cofactor in the electron transfer between Q_A and Q_B (Allen et al., 1988; MacMallan *et al.*, 1995).

Then, plastocyanin (Pc) will diffuse to the oxidation site of photosystem I (PSI). This sequence of reactions is known as the Z-scheme (Figure 13).

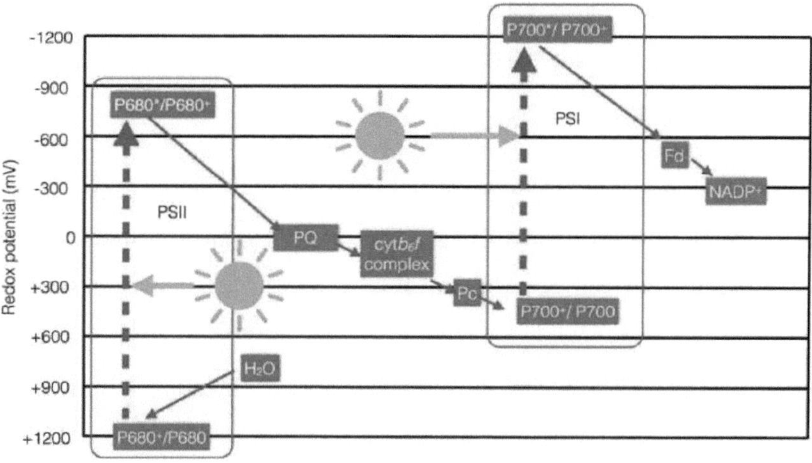

Figure 13: Z diagram of the electron transfer chain. The main constituents of the electron transfer chain are distributed according to their redox potential (ordinate). The diagram also illustrates the significant endergonic effect of light pulses on the reaction centers of PSII and PSI, causing the loss of an electron from P680 (from PS II) and P 700 (from PS I).

Courtesy of Portland Press as published in: Johnson, M.P., Photosynthesis, *Essays in Biochemistry,* 2016; 60 255-273.

PSI regulates electron transport between two mobile proteins, plastocyanin (Pc) and ferredoxin (Fd). Following light absorption by the PSI pigments, the latter become excited and transfer energy to the chlorophylls of the PSI reaction center which, in its excited state, becomes capable of rapidly transmitting an electron to its primary acceptor and finally delivering it, via other intermediates, to Fd. The latter, with the help of the enzyme ferredoxin $NADP^+$ oxidase, carries electrons to $NADP^+$ to form NADPH. The protons released during the photolysis of water and the oxidation of mobile plastoquinone (PQH_2) ensure a pH of 5.5 inside the thylakoids on the luminal side and about 8 on the stromal side. These protons will be pumped out by the ATPase which will produce ATP in the process. NADPH and ATP are metabolites necessary for the functioning of the Calvin cycle (CO_2 fixation, not shown).

These two processes PSII and PSI are coupled, Q_A transfers its electron to Q_B only when the site of Q_B is occupied by a receptor molecule, ideally the plastoquinone Q_B. Thus, if the site of Q_B is vacant as a result of the double protonation, electron transfer between Q_A and Q_B will first require the binding of a plastoquinone to Q_B and then the transfer can proceed. If the site of Q_B remains vacant, or is occupied by another molecule, there will be a change in the redox reaction of Q_A, WHICH induces fluorescence variations that can then serve as an index of integrity in the functioning of photosystems, an index that is then integrated into the development of a biosensor. In addition, the vacant sites can be occupied by structural analogues, for example certain types of herbicides, which will then induce the inhibition of electron transport. This inhibition can also be measured by fluorescence perturbations.

3.2.3 Pathways to evaluate thylakoid signal processing

Various techniques are available to measure the activity of photosystems. When chloroplasts from plants, algae or cyanobacteria are used, several parameters or metabolites can be measured (Figure 14): pH measurement, oxygen release, CO_2 consumption, ATP production, heat emission and fluorescence. These measurements are not straightforward and do not apply to all of the biosensor biological material preparations presented. Thylakoid membrane preparations may not have the entire ATPase complex or Calvin cycle molecules. To obtain a simple, fast, reliable, scientifically proven assay with commonality in the evaluation of the efficiency of chlorophyll-based photosystems, a readout mode accessible to all these models must be chosen. Fluorescence is a wise choice for both living algae and cyanobacteria and for thylakoid membrane preparations from higher plants.

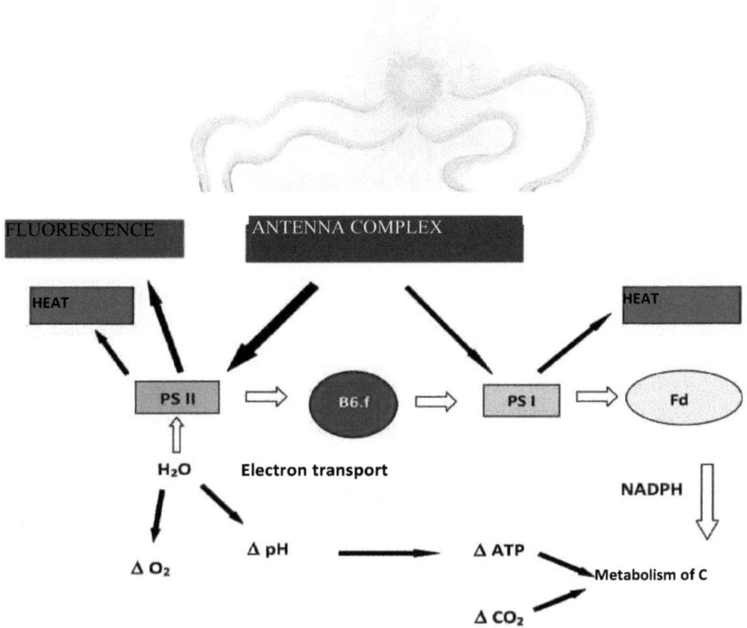

Figure 14: Schematic of light energy dissipation. Following the absorption of photons by the antenna complexes of the thylakoid membrane, PSII uses water (H_2O) to hydrolyze it into oxygen (O_2) and protons, the process releases electrons which, via a cascade, are transferred (white arrows) to cytochrome b6/f (b6/f) and

PSI and finally to ferredoxin (Fd). The hydrolysis of water induces the acidification of the medium (pH) because of the protons it releases. The pH gradient formed on both sides of the thylakoid membrane, will activate an ATPase, which will produce ATP (energy source for the cell), necessary for the metabolism of carbon (C) as well as the NADPH also produced. In addition to the production of chemicals, the excess energy received from photons and not used by electron transport will be dissipated by the movement of molecules, the release of heat and the release of photons (fluorescence). However, as collateral effects, free radicals, singlet oxygen and other oxidative substances are produced. The various pigments also serve to protect the photosystems against these radicals.

3.2.4 The fate of photons in photosystems as observed by fluorescence

4   The transfer of excitation energy (photons) captured by the antennae results in the excitation of P680, a Chl a pair located between the $_{DI}$ and $_{D2}$ subunits (Figures 8 and 9). After the formation of $_{P680*}$, the energy of P680* returns to the PSII antennae where charge separation is initiated. In this case there is reduction of pheophytin (Ph) and oxidation of P680, the P680+/Ph couple$^-$ is formed. This state can be deactivated in 4 ways:

1.   transfer of electrons from Ph to $_{QA}$ (reduction of $_{QA}$)

2.   recombination and repopulation of P680*.

3.   triplet state training

4.   various non-radiative deactivation patterns

The deactivation mode involving the transfer of electrons from Ph to $_{QA}$ was modeled by Duysens and Sweers (1963). They proposed a model involving the following molecules: Ph, $_{QA}$ and $_{QB}$ and the reduction of PQ to $_{PQH2}$ (yellow rectangle in Figure 8) according to the overall PSII operating reaction:

$$2H_2O + 4 \text{ photons} + 2PQ + 4H^+ \longrightarrow O2 + 4H^+ + 2\ PQH_2 \quad (3)$$
$$\text{open centers} \qquad\qquad \text{closed centers}$$
$$\text{(plastoquinone)} \qquad\qquad \text{(plastoquinol)}$$

As already described, the overall reaction is in fact a series of redox reactions of molecules collaborating in the electron transfer chain.

In 1963, Duysens and Sweers, linked the reduction reaction (3) of the $_{PQ/PQH2}$ couple to the fluorescence emitted by the photosystems. Their hypothesis links the emission of the variable fluorescence, Fv, of PS II to the PQ/PQH_2 ratio, thus to the electron transfer reaction. They defined important parameters to explain the mechanism, with the system moving from a minimum fluorescence emission corresponding to a maximum concentration of PQ, implying that the reaction center is available to receive two electrons to react and form PQH_2. If the entire quinone pool is in the form of PQH_2, the reduction will be blocked and the photon energy from the antennae will be re-emitted as fluorescence. The transition from minimum fluorescence Fo, to maximum fluorescence, Fm is an index of the PQ/PQH_2 ratio. If all the reaction centers (RC) of PSII whose fluorescence is measured are in reduced form (PQ), they are then called open state centers. In this case, the fluorescence is at a minimum and is defined as Fo. On the other hand, the reaction centers are in the closed state when the plasquinones are in the PQH_2 form. In this case, the fluorescence reaches its maximum level and is defined as Fm. The photosystems II work between these two extremes, with an optimal PQ/PQH_2 ratio. Fm', intermediate value of fluorescence between Fo and Fm, is an index of the PQ/PQH_2 ratio.

3.2.5 Fast fluorescence measurements, measurements of the effect of photons on PS II

Using fast kinetic fluorescence measurements, Strasser and Govindjee (1991) obtained curves identifying the intermediates suggested by the classical model of Duysens and Sweers (1963). The intermediate states of fluorescence intensity are a function of the respective open CR/closed CR ratios. These steps in the electron transfer reaction between Ph and PQH_2 are associated with the OJIP fluorescence measurements as measured in fast kinetics (Figure 15), according to the experimental conditions of Strasser and Govindjee (1991).

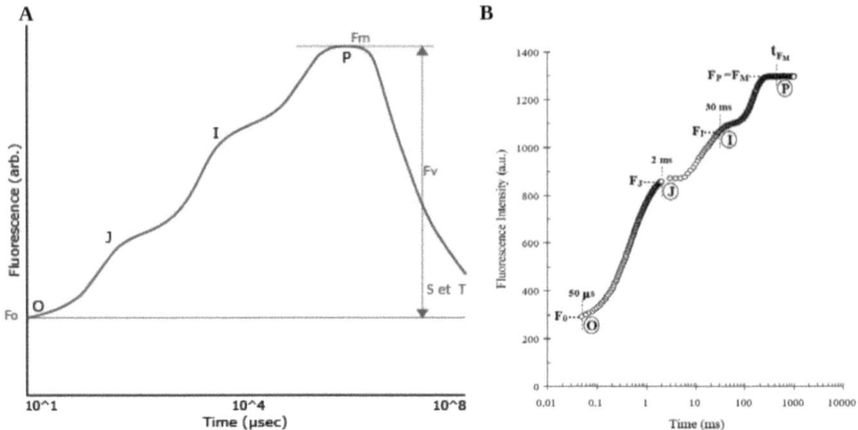

Figure 15: Simulation of the fluorescence signal decay after reaching F_m . The decay is arbitrarily identified in the plot A, by the points S and T. In B, conventional plot of photosystems fluorescence measurement as obtained on a PEA type fluorometer. This is the O-J-I-P plot: these points are associated with the equilibria between the different intermediates of the redox reaction PQ/PQH_2 as presented in Table 2. The experimental conditions for measuring fluorescence are such that F_o is 50 µsec, F_J is 2 msec, and F_I is 30 msec; the maximum fluorescence intensity is indicated by F_p and corresponds to F_m. From: Fluorescence Transient Analysis - Courtesy of Hansatech Instruments. www.hansatech-instruments.com/wp-content/uploads/2013/07/anal

When PQH_2 detaches and diffuses into the membrane, the fluorescence decreases from the P level to arbitrary S and T levels (shown in Figure 15 A), and the cycle continues. The list of intermediates between Ph and Q_B proposed by Strasser et al. (2000) is described in Table 2. These intermediates, are inspired by fast kinetic measurements obtained using PEA technology (Hansatech, Figure 15 B). The decrease in fluorescence to the S and T levels reflects the redirection of PQH_2 to the cytochrome b complex$_6$ /f to transfer protons to it (Figure 15 A).

Table 2. Intermediates according to the reaction mechanism as proposed by Strasser et al. (2000)

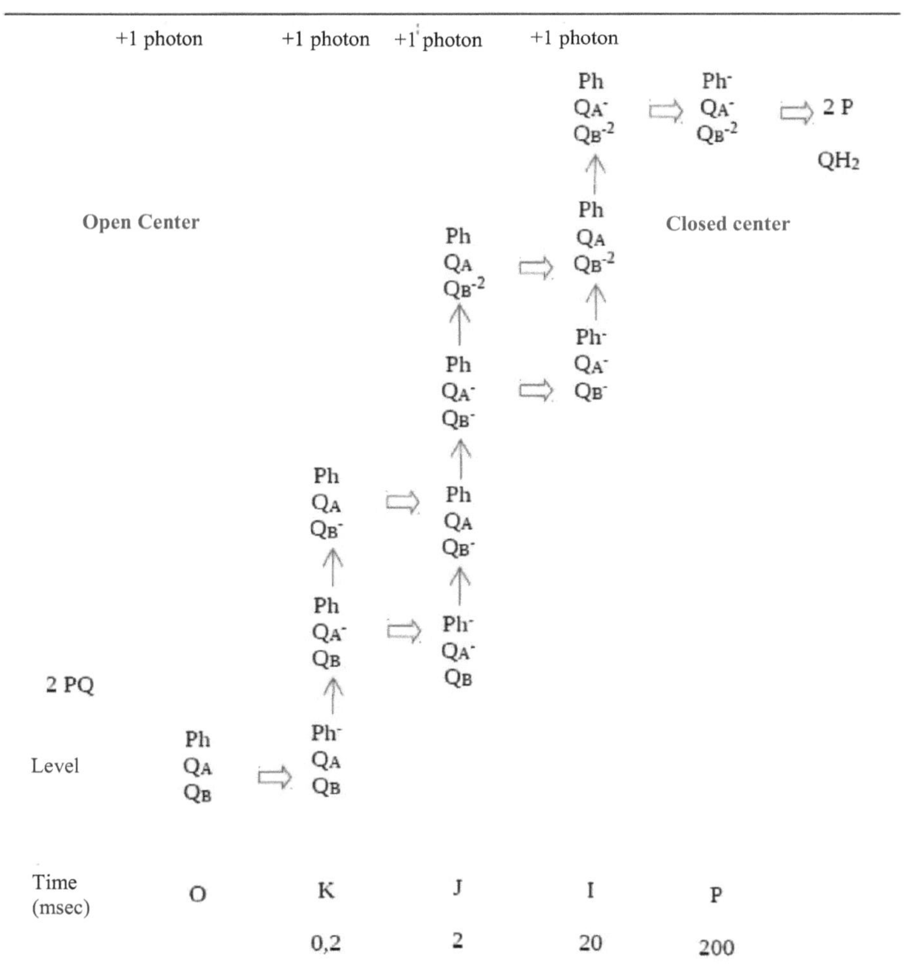

The cycle continues and the QB site then accepts a new PQ molecule available for another double protonation.

3.2.5 Integrity of photosystem II as monitored by fluorescence

In another typical protocol for measuring the activity of qualified Kautsky effect photosystems, a light flash is sent onto a surface containing PS II to follow their activation and relaxation by the fluorescence measured as a function of time. Here is what happens: the electron acceptors of the PSII become saturated. The light energy exceeding the collection possibilities of the PSII is then re-emitted in the form of fluorescence, according to a particular kinetic. Then, the electron transfer chain starts and reaches a steady state. By measuring the fluorescence, during a photon pulse, the intensity of the fluorescence rises rapidly from an initial value indicated Fo to a maximum value (Fm) in less than a second. As it is a pulsed light (flash) that saturates the photosystems, the fluorescence curves, when taken periodically following a new flash, decrease in amplitude, showing the exhaustion of the photosystems. The fluorescence signal comes mainly from the PSII. This technology is marketed under the name of PAM (pulse amplication modulation) (Schreiber et al, 1986), a typical result of which is shown in Figure 16.

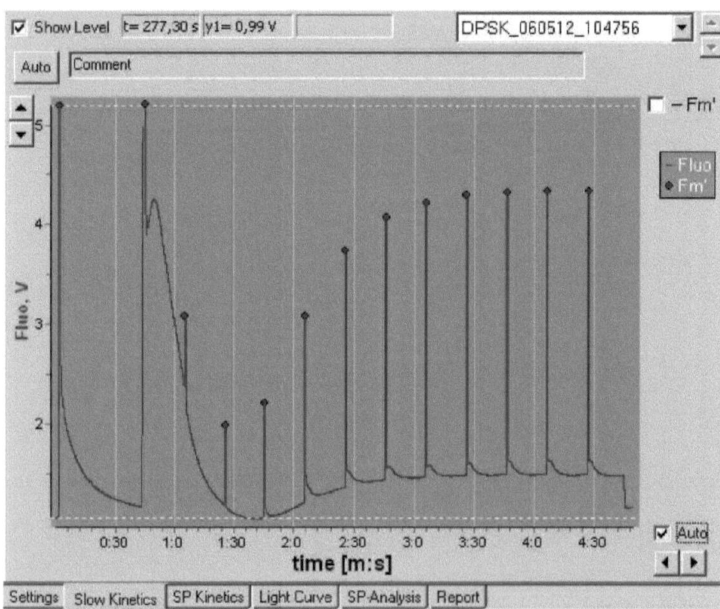

Figure 16: Fluorescence relaxation kinetics as generated by the PAM fluorometer (Heinz Walz GmbH, Germany). Each peak corresponds to a light pulse that causes the PSII to produce maximum fluorescence (Fm) and a descending curve corresponds to the relaxation of the PSII. A sequence of light pulses every 20

sec (in this example) induces a lower fluorescence measurement (Fm') due to photobleaching of the PSII. Fo is induced by low intensity modulated light, this is the lowest fluorescence level (Courtesy of Stephen Walz.)

Two types of information can be derived from these curves:

Firstly, the identification of remarkable points that evolve according to the nature and degree of stress suffered by the PS II and secondly the parameterization of the curves from these points in order to quantify the stress suffered and to interpret physiologically the phenomena observed. These points or parameters are :

- Fo or fluorescence at initial state: fluorescence intensity when all reaction centers of the PSII are open (oxidized quinones). This is the minimum fluorescence obtained when the PS II are dark adapted;

- Fm or maximum fluorescence: fluorescence intensity when all reaction centers of the PSII are closed. It is the maximum fluorescence in the dark adapted state;

Between these extremes multiple intermediate or variable physiological conditions may exist. These intermediate conditions of PSII, or even of the plants, can be measured by the measures defined, among others, by the following parameters:

- Fv/Fm: this is the ratio of variable fluorescence, Fv is defined as the fluorescence between Fm and Fo and quantitatively (Fm - Fo). This value is proportional to the photochemistry quantum yield and shows a high degree of correlation with the quantum yield of net photosynthesis. It is representative of the efficiency of PSII in using light for photochemical conversion. Its value is about 0.8 in a healthy plant, and decreases under stress. This parameter is widely used to monitor the effect of environmental stresses *in situ*.

- TFm or time of maximum fluorescence: it is the time necessary to obtain Fm. The more electron transfer is blocked at PSII, the shorter the time to obtain Fm. This parameter then becomes an indication of a decrease in the number of healthy photosystems available;

- Surface: it is the surface above the fluorescence curve between Fo and Fm. In general, this parameter is divided by ((F_{m-Fo}) or Fv); it thus expresses the energy necessary to close all the reaction centers. In case of stress, this value falls sharply;

- F(x): Fluorescence value at a given time.

As can be seen, a set of parameters useful for characterizing the efficiency of PS IIs can be calculated. They are an assessment of the ability of PSII to trap light and utilize it. With this in mind, to assess the modulation of PS II as read by PS II fluorescence, we designed a test to measure the presence of contaminants in water. The device estimates the state of PS II by fluorescence and displays the results following a summary calculation based on the Fv/Fm reading.

3.3 Biosensor: The transducer or reader sensitive to the biological signal measured: an adapted and programmed fluorometer

The biosignal sensitive transducer is a fluorometer adapted and programmed to assess the efficiency of PS II, measuring the equivalent of the Fv/Fm parameter. This is a direct measurement that has been validated (Dewez et al., 2007) and requires no reagents other than photons and photosynthetic systems either in the form of thylakoid membranes from plants, algae or cyanobacteria. The efficiency of PS II is measured via the excess fluorescence, at lengths above 700 nm. Beforehand, a light flash is sent to the reaction centers: 450 nm for chlorophyceae and 605 nm for cyanobacteria. An incubation of 15 minutes is sufficient and corresponds to the time necessary for the photosystem to adjust in the medium. This means that within 15 minutes an impact response, such as the presence of certain contaminants (Figure 17) that inhibit or destroy the photosystems, will result in a modulation of the efficiency of PS II.

3.3.1 Speed of the biosensor vs. natural protection mechanisms

Like any organized living system, algae, cyanobacteria and photosynthetic systems have defense mechanisms that protect them from destructive agents. For example, if the destructive agent is a metal cation, photosynthetic organisms of eukaryotes produce peptides that bind metal cations. These molecules, organometallic complexes, are distributed in vacuoles to allow cytoplasmic control of the concentrations of these cations, thus neutralizing their toxic potential (Cobbett et al., 2002; Perales-Vela et al., 2006). However, this mechanism takes a few hours to set up. If the measurement is made in less than an hour, i.e. before the protective system is set up, it will indicate whether the incubation environment is disturbed. If instead of algae and cyanobacteria, extracts of thylakoid membranes are used, a direct measurement of the effect of the contaminants is obtained, because no vacuolation-type protection system is active.

3.3.2 Breakdown of the integrity of photosystems II by inhibitors or contaminants

Figure 17 illustrates the potential sites of action of various contaminants or families of contaminants most commonly found in industrial or municipal wastewater. Table 3 details these sites of inhibition as well as the contaminants that can be detected by any biosensor based on the inhibition of photosynthetic activity.

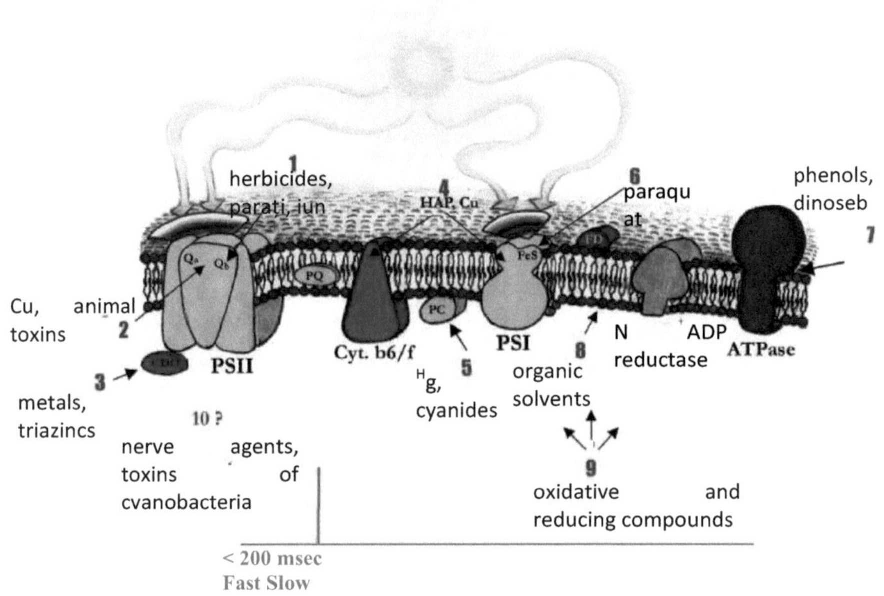

Figure 17: Sites of inhibition of contaminants on the electron transport chain. Different inhibitors act on sites in the photosynthetic apparatus. OJIP fluorescence levels are measured below 200 msec; longer times correspond to a decrease in fluorescence to the T level.

In the presence of inhibitors on thylakoid constituents (Table 3), several responses on the part of PS II can be observed. However, what is measured is akin to a decrease in the number of integral PS II in the sample. This translates into a real decrease in the photosynthetic efficiency of the tested sample: F_o remains the same, but F_m is decreased, it is called Fm·, indicating that under the new conditions, because there is inactivation of PS II, the F_m value of the control, uninhibited sample will not be reached.

Table 3: Sites of inhibition of contaminants on thylakoids

Inhibition site on the electron transport chain	Chemical contaminants	References
Site # 1: Acceptor side at the plastoquinone B (Q$_B$) site of photosystem II (PSII)	Triazines: e.g. atrazine, simazine, cyanazine Ureas: diuron, linuron etc Uracils: lenacil etc Anilides : propanil etc Phenolics: bromoxynil, bentazone etc Triazinones: metribuzin etc. Thiazoles amino and ureas Insecticide (parathion)	Boger et al. 2000; Draber et al. 1991; Kleczkowski, 1994; Prevot et al. 1993; Dayan et al. 2000; Rodriguez et al, 2002
Site # 2: Acceptor side of PSII (between Q$_A$ and Q$_B$)	Metal ions: e.g. Cu Allelochemicals and polyphenolics: e.g. tellimagrandin II	Baron et al, 1995; Mohanty et al, 1989; Renger et al, (1993); Singh et al, 1987; Yruela et al, 1993; Leu et al, 2002
Site # 3: Oxidizing side of PSII	Metal ions: (Cu, Cd, Hg, Pb, Zn)	Bernier et al, 1993; Bernier et al, 1995; Boucher et al, 1999; Maksymiec et al, 1988; Rashid et al, 1990; Rashid et al, 1991; Rashid et al, 1994; Renganathan et al, 1989, 1990; Samson et al, 1988; Schroder et al, 1994
Site # 4: Cytochrome b6/f	Polycyclic aromatic hydrocarbons (PAH); Metal ions: Cu	Brack et al, 1998; Huang et al, 1997; Mallakin et al, 2002; Singh et al, 1987
Site # 5: Plastocyanine (PC)	Cyanides Metal Ions: Hg	Berg et al, 1975; Rodriguez et al, 2002; Kimimura et al, 1972; Yocum et al, 1977; Hader et al, 1987
Site # 6: F$_B$ (PSI electron acceptor)	Herbicide (paraquat)	Fujii et al., 1990

<u>**Table 3 (continued)**</u>

Site # 7: NADP-ferredoxin-oxido reductase enzyme	Metal ions: Cu, Hg	Cedano-Maldonado et al, 1972; Shioi et al, 1978; Samuelsson et al, 1980; Honeycutt et al, 1972; DeFilippis et al, 1981
Site # 8 : Decouplers of electron transport and photophosphorylation chain	Phenols (pentachlorophenol), nitro-aromatic compounds (e.g. dinoseb)	Brack et al, 1998
Site # 9: Lipidic bilayer of the membrane	Organic solvents	Brack et al, 1998
Site # 10: CO_2 fixation (Calvin cycle)	Sulfite and bisulfite Cyanide	N'Soukpoé-Kossi et al, 1994 Jones et al. 1999
	Reactive oxidizing or reducing compounds	Brack et al. 1998,
Sites not determined	Nerve agents: tabun, sarin, mustard gas, tributylamine and dibutyl sulfide; cyanobacterin	Sanders et al. 2001 Gleason et al, 1886

4 Toxicity biosensor: standardized conditions for the evaluation of the efficacy of SP II

Similar to toxicity microtests, a biosensor sensitive to many pollutants can be made from chlorophyll photosystems because, 1) photosynthetic systems of thylakoids either plants, algae and cyanobacteria can be a good biological material in several aspects, 2) photosynthetic systems are sensitive to several molecules that can be found in water and, 3) the excess energy dissipated and measured by the fluorescence of PS II correlates with the efficiency of PS II. The PEA and PAM technologies that measure this parameter remain complicated for the uninitiated. However, they provide important parameters for the characterization of photosystems. A proper protocol for the use of biological material conjugated to a fluorometer dedicated to the measurement of apparent Fv/Fm becomes a tool accessible to the uninitiated.

If we want to use a biosensor based on these biological materials to provide a fast, simple and economical micro-test for the detection of pollutants in water, we need to standardize a protocol, a biological material and a measurement method. In order to evaluate the efficiency of the photosystems, we have privileged the approach of the initial velocities, methodology used in chemical kinetics which allows to evaluate the apparent velocity of a reaction. It is an approach that has been proven in analytical chemistry and enzymology to determine the effects of inhibitors on reaction parameters. It consists in selecting a fixed time window and evaluating the reaction rate. The parameter measured by the adapted fluorometer is the apparent photochemical yield or the efficiency of the photosystems during this time window. To simplify the interpretation of the results and to relate them to a relative but representative value, the following physico-chemical parameters are controlled: temperature, quality and concentration of photosystems (by measuring chlorophylls), reaction medium, wavelength and quantity of photons sent (intensity) to the photosystems, and measurement time. With this in mind, we have written Volume II which presents detailed protocols as well as several applications in water and sediments obtained by various laboratories.

5 Discussion: Looking to the Future

The deployment of this new science, ecotoxicology, from the point of view of its growing complexity, has led to the interest of scientists from all backgrounds. This has led to the development of ecosystem models that encompass all spheres of life, including the geographical environment: a biology-geography conjunction that has given rise to international meetings and climate change objectives. This new approach needs measurement tools, a look, an eye that allows for rapid and adequate analysis and transmission of results. These are complementary tools that must be developed and learned to use adequately.

Applied in the water sciences, tools for measuring the ecotoxicological response have been developed and have become more significant than measuring the presence of one or more contaminants; it is a look at the impacts of water quality on living organisms (Figure 5). Ecotoxicological measurements will allow biomonitoring of water by measuring the improvement or degradation of its quality. Biomonitoring depends on the processes put in place to verify and follow the evolution of the measured parameters. For some years now, on-line analysers have been used to measure important physico-chemical parameters. More complete analysis stations are beginning to be installed at strategic locations and the demand for toxicity micro-tests is growing. Research is increasing in this area, but the difficulty is that a single test is not enough. In ecotoxicology, several tests must be performed in order to target the extent of the living: plant and aquatic for the example of water. It is the crossing of the results obtained with 3 to 5 toxicity micro-tests that will provide a quality of monitoring. This suggestion has been discussed in depth by various scientists and detailed in the book published by Blaise and Férard in 2005: Hazard Assessment Schemes. It presents and discusses various cross-analyses of toxicity micro-test results obtained in various aquatic environments.

All these results, gathered and analyzed by computer programs, allow us to monitor and predict trends or even act as an alert station.

Therefore, it is important to develop and use biosensors such as the one presented in this paper. We have exploited only the transmission aspects of thylakoids and mainly the fluorescence of PS II via the measurement of the Fv/Fm parameter by the initial velocity method, it is the one that seemed to us the simplest. In Volume II we present the diversified sectors where the photosynthetic systems biosensor has been used.

It is now up to us to use and develop management tools for the decontamination of municipal and industrial wastewater in order to concretize and democratize water biomonitoring. We thus join our voice to the National Academy of Pharmacy which quotes in its report of September 2008[2] :

"Current regulations also impose a number of discharge permit controls based on concentrations of TSS, COD, BOD, total nitrogen, total phosphorus, SEC, detergents, AOX, phenols, total hydrocarbons, Al, Ag, Cd, Cr VI, total Cr, Cu, Fe, Hg, Pb, Zn, nitrites, and nitrates, most of which are not relevant to healthcare facilities. Medicines are not subject to these discharge authorizations.

Strengthen environmental monitoring of discharges from the chemical and pharmaceutical industries, health care facilities, industrial and fish farms, and all activities that may cause discharges of medicinal substances or their residues, and ***improve the treatment of these point source discharges.***

*Take into account the effects related to the multiplicity of substances present in discharges by developing **global toxicity tests**, in particular for carcinogenic, mutagenic and reproductive toxic substances."*

The better the water is monitored, the better the quality. It's up to all of us now to take action to improve water biomonitoring and leave a healthy environment for future generations.

2 Medicines and the Environment: Report of the National Academy of Pharmacy, Sept 2008.

6 Bibliography

Adir N. (2005). Elucidation of the molecular structures of components of the phycobilisome: reconstructing a giant, *Photosynth. Res.* 85: 15-32.

Allen J.P. , Feher G., Teates T.0., Komiyas H., Rees D.C. (1988). Structure of the reaction center from Rhodobacter sphaeroides R-26: Protein-cofactor (quinones and Fe21) interactions. *Proc. Nati. Acad. Sci. USA* (85), 8487-8491

Barber J. (1993). Photosynthetic Oxygen Evolution. Bioenergetics Group and British Photobiology Society Joint Colloquium, Edited by M. C. W. Evans. 649th Meeting held at Imperial College, London, 19-21 Dec, 313-318.

Baron M., Arellano J.A., Gorgé J.L. (1995). Copper and photosystem II: A controversial relationship. *Physiol Plant.* 94: 174-180.

Berg S.P. and Krogmann D.W. (1975). Mechanism of KCN Inhibition of Photosystem I. *J. Biol. Chem.* 250 (23): 8957-8962.

Bernier M., Popovic R., Carpentier R. (1993). Mercury inhibition of photosystem II by chloride. *FEBS Lett.* 321: 19-23.

Bernier M. and Carpentier R. (1995). The action of mercury on the binding of extrinsic polypeptides associated with the water oxidizing complex of photosystem II. *FEBS Lett.* 360: 251-254.

Blaise C., (2000). Canadian application of microbiotests to assess the toxic potential of complex liquid and solid media. In: *New microbiotests for routine toxicity screening and biomonitoring*, Kluwer Academic/Plenum Publisher, New York, 550 p: 3-13.

Blaise C., Férard J.F.,(2005). Small-Scale Freshwater Toxicity Investigation. Ed. by Christian Blaise and Jean-François Férard. Published by Springer, Volume I.

Blaise C., Férard J.F.,(2005). Hazard Assessment Schemes. Ed. by Christian Blaise and Jean- François Férard. Published by Springer, Volume II.

Boger P. and Sandmann G. (2000). Action of modern herbicides. In *Photosynthesis.* A Comprehensive Treatise. Raghavendra A.S. (ed.) Cambridge University Press, New York, pp. 337-351.

Boucher N. and Carpentier R. (1999). Hg^{2+} , Cu^{2+} and Pb^{2+} induced changes in Photosystem II photochemical yield and energy storage in isolated thylakoid membranes: A study using simultaneous fluorescence and photoacoustic measurements. *Photosynth. Res.* 59: 164-174.

Brack W, Frank H. (1998).Chlorophyll a fluorescence: a tool for the investigation of toxic effects in the photosynthetic apparatus. *Ecotoxicol Environ Saf.* 40(1-2):34-41.Bryant, D. A., G.

Guglielmi, N. Tandeau de Marsac, A. Castets and G. Cohen-Bazire (1979). "The structure of cyanobacterial phycobilisomes: a model". *Arch Microbiol* **123**: 113-127.

Cairns John Jr, (2005). Biomonitoring: The crucial link between Natural systems and society, *Mankind Quaterly*, Volume XLV, number 3, 2205, 289-308).

Cedeno-Maldonado A., Swader J.A., Heath R.L. (1972). The cupper ion as an inhibitor of photosynthetic electron transport in isolated chloroplasts. *Plant Physiol.* 50: 698-701.

Cobbett, C., Goldsbrough, P., 2002. Phytochelatin and metallothioneins: Roles in heavy metal detoxification and homeostasis. *Annu. Rev. Plant Biol.* 53: 159-182.

Conrad R., Büchel C., Wilhelm C., Arselane W., Berkaloff C., Duval J.-C. (1993). Changes in yield of *in vivo* fluorescence of chlorophyll *a* as a toll for sensitive monitoring. *J, Appl. Phycol.* 5: 505516.

Croce R. and van Amerongen H. (2014). Natural strategies for photosynthetic light harvesting. *Nature chemical biology*, 10: 492 -501.

Dau H. (1994a). Molecular mechanisms and quantitative models of variable photosystem II fluorescence. *Photochem. Photobiol.* 60(1): 1-23.

Dau H. (1994b).Short-term adaptation of plants to changing light intensities and its relation to Photosystem II photochemistry and fluorescence emission. *J. Photochem. Photobiol. B Biol.* 327.

Dayan F.E., Vincent A.C., Romagni J.G., Allen S. N., Duke S.O., Duke M.V., Bowling J.J., Zjawiony J.K. (2000). Amino- and Urea-Substituted Thiazoles Inhibit Photosynthetic Electron Transfer. *J. Agric. Food Chem.* 48: 3689-3693.

de Filippis L.F., Hampp R., Ziegler H. (1981). The effects of sublethal concentrations of zinc, cadmium and mercury on *Euglana. Arch. Microbiol.* 128: 407-411.

Dewez D., Marchand M., Ellaffroy P., Popovic R. (2002). Evaluation of the Effects of Diuron and Its derivatives on Lemna gibba Using a Fluorescence Toxicity Index. *Environ. Toxicol.* 17: 493501.

Dewez D. Boucher N., Bellemare F. and Popovic R. (2007) Investigation of measurement sensitivity by using LuminoTox, PAM and PEA fluorometric systems when PSII thylakoid membrane was exposed to atrazine and copper effects : *Toxicol. Env. Chem.* 89 : 655-664.

Draber W., Tietjen K., Kluth J. F., Trebst A. (1991). Herbicides in Photosynthesis Research. *Angew. Chem.* 30: 1621-1633.

Duysens LNM, Sweers HE (1963). Mechanism of two photochemical reactions in algae as studied by means of fluorescence. In: *Japanese Society of Plant Physiologists* (eds) Studies on Microalgae and Photosynthetic Bacteria, pp 353-372. University of Tokyo Press, Tokyo.

Foyer CH (1984). Effects of CO2 on Plants: CO2 and Plants. Science. 224(4647):380-1.

Fujii T., Yokoyama E.I., Inoue K., Sakari H. (1990). The sites of electron donation of photosystem I to methyl viologen. *Biochem. Biophys. Acta* 1015: 41-48.

Ghanotakis D. F. and Yocum C. F. (1990). Photosystem II and the oxygen-evolving complex. *Ann. Rev. Plant Physiol. plant Mol. Biol.* 41: 255-276.

Glazer A.N. (1984). Phycobilisome - a macromolecular complex optimized for light energytransfer, *Biochim. Biophys. Acta* 768: 29-51.

Gleason F.K. and Case D.E. (1986). Activity of the Natural Algicide, Cyanobacterin, on Angiosperms. *Plant Physiol.* 80: 834-837.

Grenn, B. R., J. M. Anderson and W. W. Parson (2003). Photosynthetic membranes and their lightharvesting antennas. *Light-harvesting antennas in photosynthesis.* B. Grenn and W. Parson *(ed.)*, Printed in the Netherlands: 1-28.

Grossman A.R. , M.R. Schaefer, G.G. Chiang, J.L. Collier (1993) The phycobilisome, a light harvesting complex responsive to environmental-conditions, *Microbiol. Rev.* 57: 725-749.

Havaux, M, Niyogi, K.K. (1999). The violaxanthin cycle protects plants from photooxidative damage by more than one mechanism. *Proc. Natl. Acad. Sci. USA* Vol. 96, pp. 8762-8767

Hader D.P. and Tevini M. (1987). *General Photobiology.* Pergamon Press, Oxford.

Honeycutt R.C. and Krogmann D.W. (1972). Inhibition of chloroplast reactions with phenylmercury acetate. *Plant Physiol.* 49: 376-380.

Hsu B.D., Lee J.Y., Pan R.L. (1986). The two binding sites for DCMU in photosystem II. *Biochem. Biophys. Res. Com.* 141 (2): 682-688.

Huang X.-D., McConkey B.J., Babu T.S., Greenberg B. M. (1997). Mechanisms of photoinduced toxicity of photomodified anthracene to plants: inhibition of photosynthesis in the aquatic higher plant *Lemna gibba* (duckweed). *Environ. Toxicol. Chem.* 16 (8): 1707-1715.

Hugo Virgilio Perales-Vela, Julia'n Mario Pena-Castro, Rosa Olivia Canizares-Villanueva (2006). Heavy metal detoxification in eukaryotic microalgae. *Chemosphere* 64: 1-10

Isomaa, B. and Lilius H. (1995). The Urgent Need for *In Vitro* Tests in Ecology. Toxic *in Vitro.* 9 (6): 821-825.

Iwasaki I., Tamura N., Okayama S. (1995). Effects of Light Stress on Redox Potentials Forms of Cyt. b559 in Photosystem II Membranes Depleted of Water Oxidazing Complex. *Plant Cell Physiol.* 36: 583-589.

Jones R.J. and Hoegh-Guldberg O. (1999). Effects of cyanide on coral photosynthesis: implications for identifying the cause of coral bleaching and for assessing the environmetal effects of cyanide fishing. *Mar. Ecol. Prog. Ser.* 177: 83-91.

Joshi M.K. and Mohanty P. (1995). Probing of photosynthetic performance by chlorophyll a fluorescence:

Analysis and interpretation of fluorescence parameters. *J. Sc. Ind. Res.* 54: 155174.

Juneau P. (2000) "Utilisation de la fluorescence chlorophyllienne des plantes comme bioessais des facteurs environnementaux", Ph.D. thesis, Montreal, Université du Québec à Montréal, 209 pages.

Keddy PA. (1994). Applications of the Hertzsprung-Russell star chart to ecology: reflections on the 21(st) birthday of Geographical Ecology. *Trends Ecol Evol.* 9: 231-4.

Kimimura M. and Katoh S. (1972). Studies on electron transport associated with photosystem I. 1. Functional site of plastocyanin; inhibitory effects of $HgCl_2$ on electron transport and plastocyanin in chloroplasts. *Biochim. Biophys. Acta* 283: 279-292.

Kleeczkowski L.A. (1994). *Ann. Plant Physiol. plant Mol. Biol.* 45: 339-361.

Krause G.H. and Weis E. (1988). The photosynthetic apparatus and chlorophyll fluorescence. An introduction. In: *K. Lichtenthaler (ed.)* Applications of Chlorophyll Fluorescence, pp. 3-11. Kluwer Academic Publishers.

Lazàr D. (1999). Chlorophyll *a* fluorescence induction. *Biochim. Biophys. Acta* 1412 : 1-28.

Leu E., Krieger-Liszkay A., Goussias C., Gross E. (2002). Polyphenolic allelochemicals from the aquatic angiosperm Myriophyllum spicatum inhibit photosystem II. *Plant Physiol.* 130: 20112018.

Liu H, Zhang H, Niedzwiedzki D.M., Prado M, He G, Gross M.L. , Blankenship R.E. (2013).Phycobilisomes Supply Excitations to Both Photosystems in a Megacomplex in Cyanobacteria. *Science*; 342(6162): 1104-1107.

MacMallan F., Lendzian F., Renger G., Lubitz W. (1995). EPR and ENDOR Investigation of the Primary Electron Acceptor Anion Q_A^- in Iron-Depleted Photosystem II Membrane Fragments. *Biochem.* 34: 8144-8156.

Maksymiec W. and Baszynski T. (1988). The effect of Cd^{2+} on the release of proteins from thylakoid membranes of tomato leaves. *Acta Soc. Bot. Pol.* 57: 465-474.

Mallakin A., Babu S., Dixon D.G., Greenberg B.M. (2002). Sites of toxicity of Specific Photooxidation Products of Anthracene to the Higher Plants: Inhibition of Photosynthetic Activity and Electron Transport Chain in *Lemna gibba* L. G-3 (Duckweed). *Environ. Toxicol.* 17: 462-471.

MacColl , R. (1998). Cyanobacterial phycobilisomes. *J. Struct. Biol.* 124: 311-334.

Medicines and the Environment: *Report of the National Academy of Pharmacy*, Sept 2008.

Mohanty N., Vass I., Demeter S. (1989). Copper toxicity affects photosystem II electron transport at the secondary quinone acceptor, Q_B . *Plant Physiol.* 90: 175-179.

Murata N. and Miyao M. (1989). Photosystem II and oxygen evolution. *Photosynthesis,* Alan R. Liss, Inc. 59-70.

Murphy DJ (1986). Reconstitution of energy transfer and electron transfer between solubilised pigment-protein complexes from thylakoid membranes. The role of acyl lipids. *Photosynth Res.* 8: 219-33

N'Soukpoé-Kossi C.N., Veeranjaneyulu K., Leblanc R.M. (1994). Sites of action of sulphite and bisulphite in the photosynthetic apparatus of sugar maple leaves as studied by photo-acoustic and modulated fluorescence methods. *Plant Cell Environ.* 17: 731-738.

Neilson J.A.D., Durnford D. G. (2010). Structural and functional diversification of the lightharvesting complexes in photosynthetic eukaryotes. *Photosynth Res* 106:57-71.

Persoone G., Janssen C., De Coen W. (2000). Preface in: *New microbiotests for routine toxicity screening and biomonitoring*, Kluwer Academic/Plenum Publisher, New York, 550 p

Pfeil B .E., - Schoefs B.,- Spetea C. (2014). Function and evolution of channels and transporters in photosynthetic membranes. *Cell. Mol. Life Sci.* 71:979-998.

Prevot P, Perret E, Jupin H, Soyer-Gobillard MO (1993). Fluorescence induction used to measure parathion toxicity in the marine *dinoflagellate Prorocentrum micans*: a possible model for biodetection. *Ecotoxicol Environ Saf.* 25(3):360-71.

Rashid A. and Popovic R. (1990). Protective role of $CaCl_2$ againts Pb^{2+} inhibition in Photosystem II. *FEBS Lett.* 271: 181-184.

Rashid A., Bernier M., Pazdernick, Carpentier R. (1991). Interaction of Zn^{2+} with the donor side of Photosystem II. *Photosynth. Res.* 30: 123-130.

Rashid A., Camm E.L., Ekramoddoullah K.M. (1994). Molecular mechanism of action of Pb^{2+} and Zn^{2+} on water oxidizing complex of photosystem II. *FEBS Lett.* 350: 296-298.

Renganathan M. and Bose S. (1989). Inhibition of primary photochemistry of photosystem II by copper in isolated pea chloroplasts. *Biochim. Biophys. Acta* 974: 247-253.

Renganathan M. and Bose S. (1990). Inhibition of photosystem II bu Cu^{2+} ion. Choice of buffer and reagent is critical. *Photosynth. Res.* 23: 95-99.

Renger G., Gleiter H., Haag E., Reifarth F. (1993). Photosystem II: thermodynamics and kinetics of electron transport from QA" to QB ($_{QB}^-$) and deleterious effects of copper (II). *Z. Naturforsch.* 48c: 234-240.

Rodriguez M Jr, Sanders CA, Greenbaum E. (2002). Biosensors for rapid monitoring of primarysource drinking water using naturally occurringphotosynthesis. *Biosens. Bioelectron.* 17: 843-9.

Samson G., Morissette J.C., Popovic R. (1988). Copper quenching of the variable fluorescence in *Dunaliella tertiolecta*. New evidence for copper inhibition effect on PSII photochemistry. *Photochem. Photobiol.* 48: 329-332.

Samson G. (1989). "Modifications of electron transport in photosystem II inhibited by mercury and copper", Ph.D. thesis, Trois-Rivières, Université du Québec à Trois-Rivières, 138 pages.

Samuelsson G. and Oquist G. (1980). Effects of copper on photosynthetic electron transport and chlorophyll-protein complexes of *Spinacia oleracea. Plant Cell Physiol.* 21: 445-454.

Sanders CA, Rodriguez M Jr, Greenbaum E. (2001). Stand-off tissue-based biosensors for the detection of chemical warfare agents using photosynthetic fluorescence induction. *Biosens. Bioelectron.* 16:439-46.

Schreiber U. Schliwa U., Bilger, W. (1986) Continuous recording of photochemical and nonphotochemical chlorophyll fluorescence quenching with a new type of modulation fluorometer. *Photosynthesis Research,* 10: 51-62.

Schroder W.P., Arellano J.B., Bittner T., Baron M., Eckert H.-J., renger G. (1994). Flash-induced absorption spectroscopy studies of copper interaction with photosystem II in higher plants. *J. Biol. Chem.* 269: 32 865-870.

Shioi Y., Tamai H., Sasa T. (1978). Effects of copper on photosynthetic electron transport systems in spinach chloroplasts. *Plant Cell Physiol.* 19: 203-209.

Singh D.P. and Singh S.P. (1987). Action of Heavy Metals on Hill Activity and O_2 Evolution in *Anacystis nidulans. Plant Physiol.* 83: 12-14.

Strasser R.J. and Govindjee (1991). The F_o and the O-J-I-P fluorescence rise in higher plants and algae. In: *J.H. Argyroudi-Akoyunoglou (ed.)* Regulation of Chloroplast Biogenesis, pp. 423-426, Plenum Press, New-York.

Strasser J. and Govindjee (1992). On the O-J-I-P fluorescence transient in leaves and D1 mutants of *Chlamydomonas reinhardtii.* In: *N. Murata (ed.)* Research in Photosynthesis, vol. II, pp. 29-32, Kluwer Academic Publishers, Netherlands.

Strasser RJ, Srivastava A and Tsimilli-Michael M (2000). The fluorescence transient as a tool to characterize and screen photosynthetic samples. In: *Yunus M, Pathre U and Mohanty P (eds)* Probing Photosynthesis: Mechanism, Regulation and Adaptation, Chapter 25, pp 443-480. Taylor and Francis, London, UK.

Tandeau de Marsac N. (2003). Phycobiliproteins and phycobilisomes: the early observations, *Photosynth. Res.* 76: 197-205

Weis E. and Lechtenberg D. (1989). Fluorescence analysis during steady-state photosynthesis. *Phil. Trans. R. Soc. Lond. B* 323: 253-268.

Xu Q. And Bricker T.M (1992). Structural Organization of Proteins on the Oxidizing Side of Photosystemm II. J. Biol. Chem. 267: 25816-25821.

Yamamoto Y. (1989). Molecular Organisation of Oxygen-Evolution System in Chloroplast. Bot. Mag. Tokyo. 102: 565-582.

Yocum F. Y. and Guikema J.A. (1977). Photophosphorylation Action associated with photosystem II cyclic photophosphorylation catalyzed by p-phenylenediamine. *Plant Physiol.* 49: 33-37.

Yruela I., Alfonso M., Ortiz de Zarate I., Montoya G., Picorel R. (1993). Precise Location of the Cu(II)-inhibitory Binding Site in Higher Plant and Bacterial Photosynthetic Reaction Centers as Probed by Light-induced Absorption Changes. *J. Biol. Chem.* 268: 1684-1689.

Zilinskas, B. A. and L. S. Greenwald (1986). Phycobilisome structure and function. *Photosynth Res* 10: 7-35.

55